FLOATING AND SINKING

FLOATING AND SINKING

By FRANKLYN M. BRANLEY

Illustrated by ROBERT GALSTER

Thomas Y. Crowell Company · New York

LET'S-READ-AND-FIND-OUT SCIENCE BOOKS

Editors: *DR. ROMA GANS*, Professor Emeritus of Childhood Education, Teachers College, Columbia University
DR. FRANKLYN M. BRANLEY, Chairman and Astronomer of The American Museum–Hayden Planetarium

Air Is All Around You
Animals in Winter
A Baby Starts to Grow
Bees and Beelines
Before You Were a Baby
The Big Dipper
Big Tracks, Little Tracks
Birds at Night
Birds Eat and Eat and Eat
The Blue Whale
The Bottom of the Sea
The Clean Brook
Down Come the Leaves
A Drop of Blood
Ducks Don't Get Wet
The Emperor Penguins
Find Out by Touching
Fireflies in the Night
Flash, Crash, Rumble, and Roll
Floating and Sinking
Follow Your Nose
Glaciers
Gravity Is a Mystery

Hear Your Heart
High Sounds, Low Sounds
How a Seed Grows
How Many Teeth?
How You Talk
Hummingbirds in the Garden
Icebergs
In the Night
It's Nesting Time
Ladybug, Ladybug, Fly Away Home
The Listening Walk
*Look at Your Eyes**
A Map Is a Picture
The Moon Seems to Change
My Five Senses
My Hands
My Visit to the Dinosaurs
North, South, East, and West
Rain and Hail
Rockets and Satellites
Salt
Sandpipers

Seeds by Wind and Water
Shrimps
The Skeleton Inside You
Snow Is Falling
Spider Silk
Starfish
*Straight Hair, Curly Hair**
The Sun: Our Nearest Star
The Sunlit Sea
A Tree Is a Plant
Upstairs and Downstairs
Watch Honeybees with Me
What Happens to a Hamburger
What I Like About Toads
What Makes a Shadow?
What Makes Day and Night
*What the Moon Is Like**
Where Does Your Garden Grow?
Where the Brook Begins
Why Frogs Are Wet
The Wonder of Stones
*Your Skin and Mine**

*AVAILABLE IN SPANISH

Copyright © 1967 by Franklyn M. Branley. Illustrations copyright © 1967 by Robert Galster. All rights reserved. Except for use in a review, the reproduction or utilization of this work in any form or by any electronic, mechanical, or other means, now known or hereafter invented, including photocopying and recording, and in any information storage and retrieval system is forbidden without the written permission of the publisher. Manufactured in the United States of America. Library of Congress Catalog Card No. 67-15396. Published in Canada by Fitzhenry & Whiteside Limited, Toronto.

ISBN 0-690-30917-1
0-690-30918-X (LB)

FLOATING AND SINKING

You can float on water.

So can wood and paper, feathers and dandelion seeds.
You can think of many other things that float.
You can think of things that sink, too.
Nails sink in water. So do stones, paper clips, dishes,
 pennies, forks, and spoons.
Why do some things float, and other things sink?

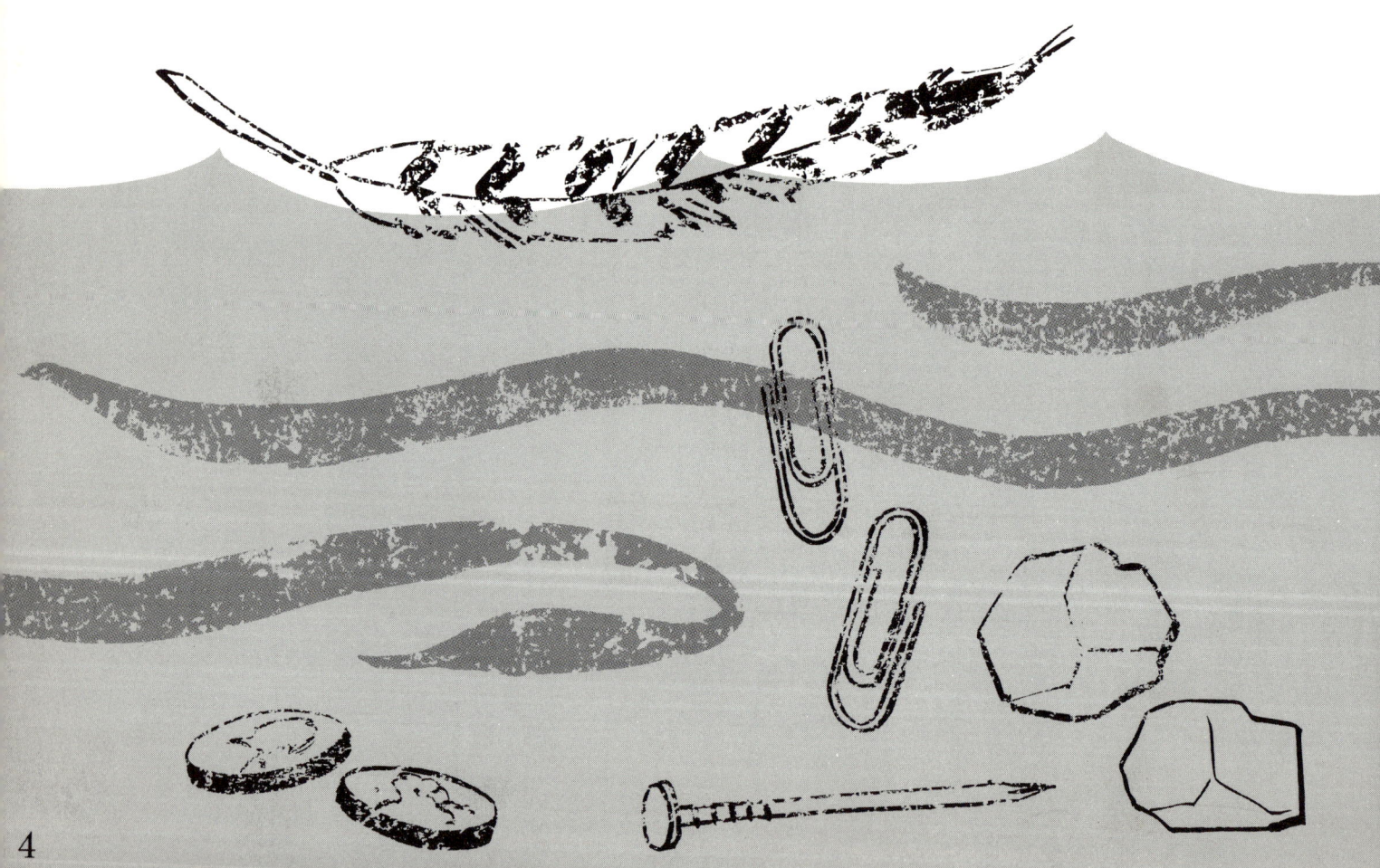

Some people say if a thing is light it floats, and if it is heavy, it sinks.
But that can't be right.
Big ships made of iron and steel are very heavy, yet they float.

Let's see if we can find out why some things float and other things sink.

Suppose we start with two small bottles of the same size and shape, a pan of water, and enough sand to fill one of the bottles. You can use salt if you have no sand.

Screw the caps on the bottles. Now you have two bottles full of air. Put the bottles in the pan of water. The bottles of air float. They sink into the water and the water in the pan rises just a little bit.

We say each bottle pushes aside a little bit of water. Suppose you weighed the water that the bottle of air pushes aside. The water would weigh the same as the bottle of air. That's why the bottle of air floats.

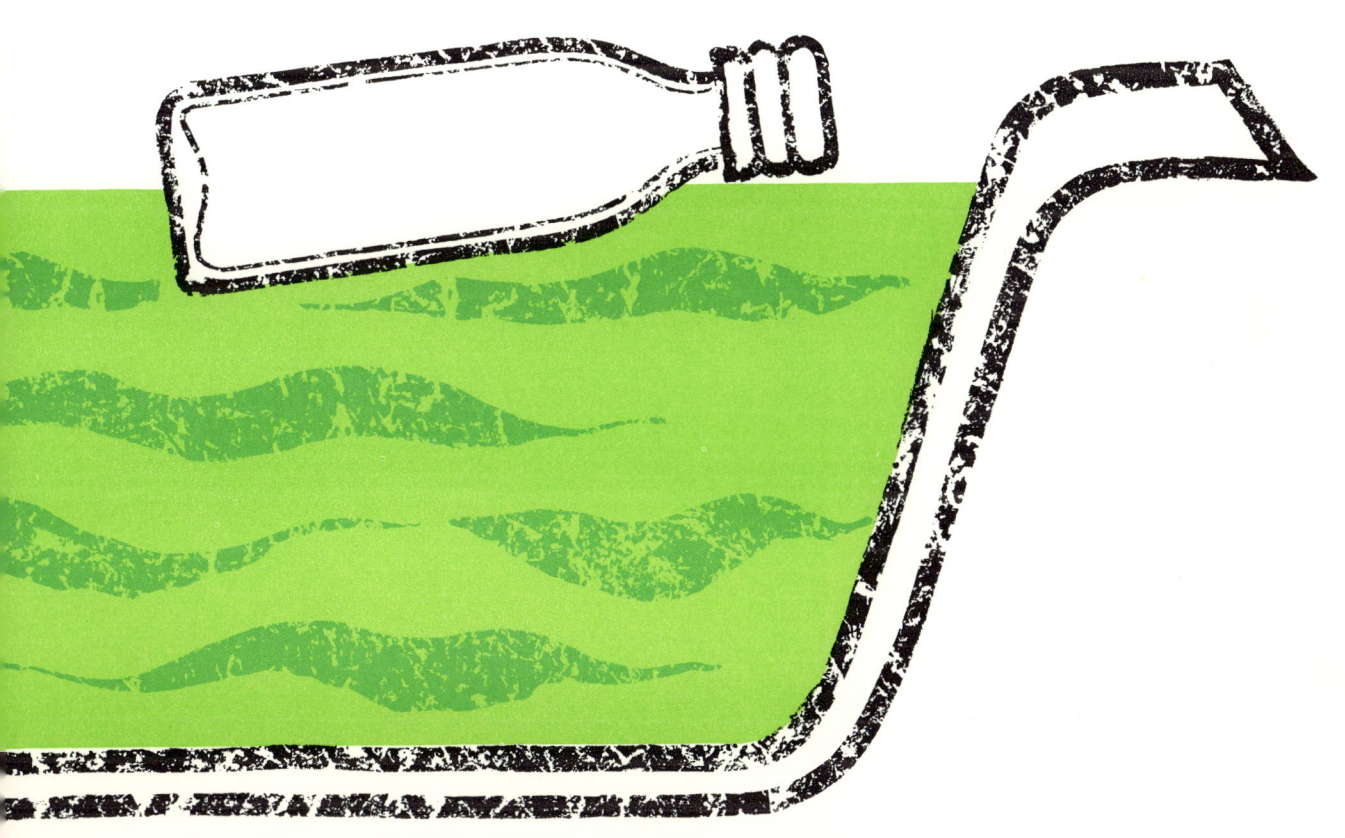

Pour a little sand into one of the bottles. Now part of the bottle is filled with sand, and the rest is filled with air. Put the cap on the bottle. Then place it in the water again. This bottle sinks deeper in the water than the bottle of air, but it still floats.

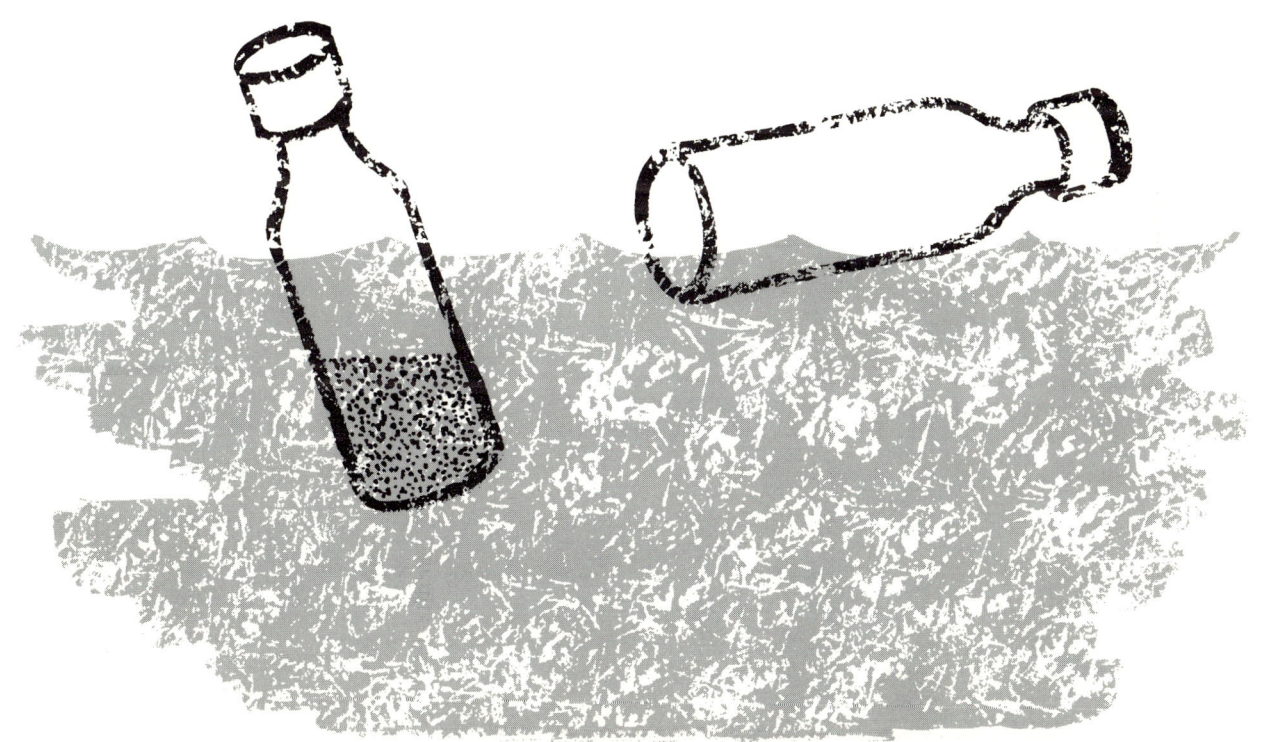

Add a little more sand. The bottle sinks a little deeper into the water.

The more sand you add, the deeper the bottle sinks. After a while, the bottle sinks all the way to the bottom.

Let's see what you did.

The bottles you used were the same size and the same shape.

You added sand to one bottle. When you added sand, you pushed air out of the bottle.

The bottle of sand weighed more than the bottle of air.

The weight of the bottle of air was the same as the weight of the water it pushed aside. That's why it floated.

The weight of the bottle of sand was more than the weight of the water it pushed aside. That's why it sank.

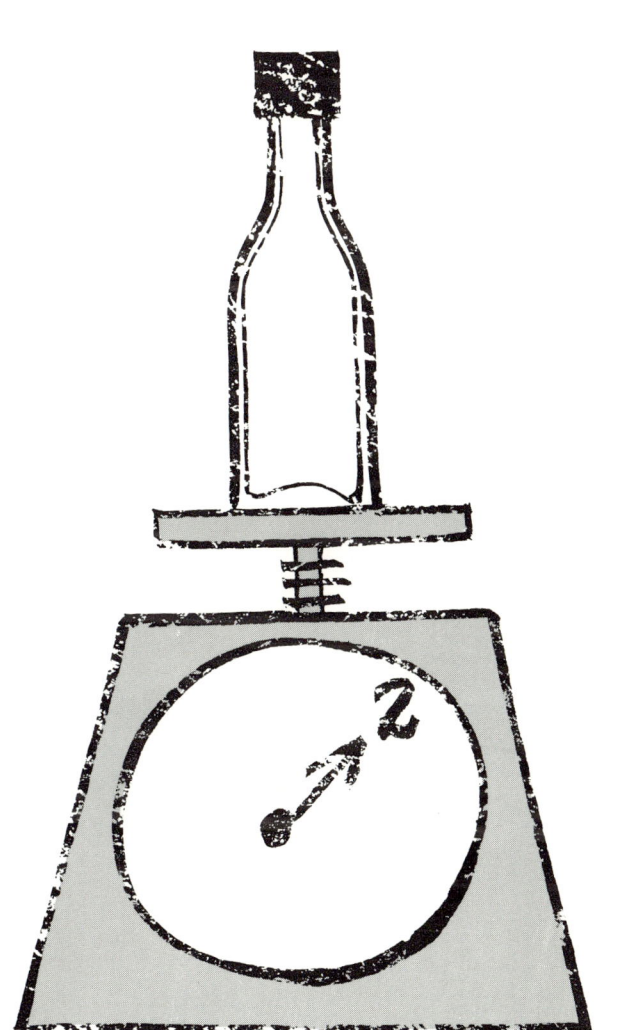 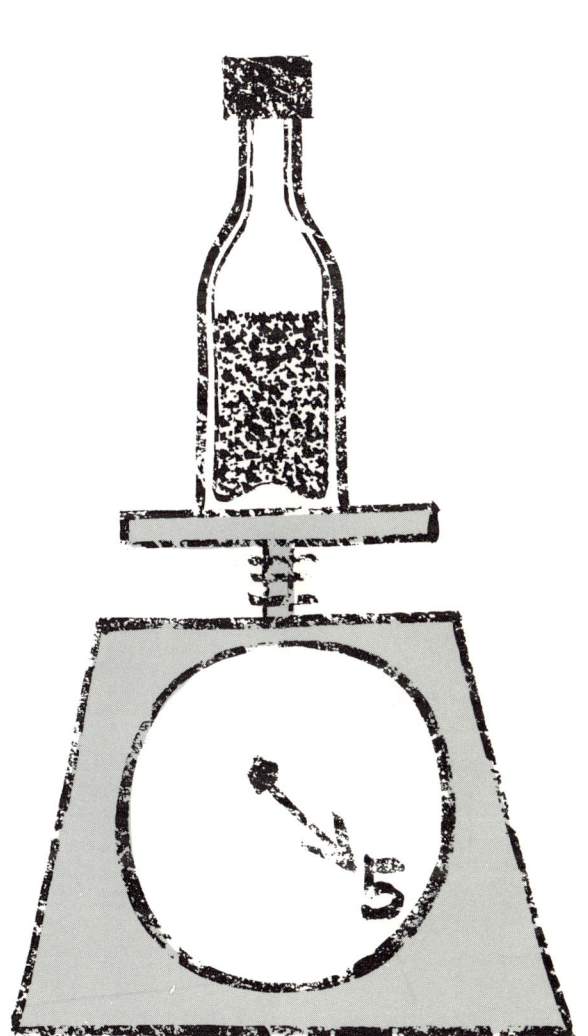

The same thing happens with a big ship.
A ship floats, but it sinks part way into the water.

The ship sinks until the weight of the water it pushes aside is the same as the weight of the ship.

When a ship carries a heavy load or cargo, it weighs more than when it is unloaded. The heavier cargo it carries, the deeper a ship sinks. It sinks until the weight of the water it pushes aside equals the weight of the ship and its load.

If a ship with its cargo weighs more than the water it pushes aside, the ship will sink.

You can see why ships must not be overloaded.

21

Submarines are ships that can float or sink. When a submarine goes under the water, we don't say it sinks. We say it submerges.

A submarine floats when the weight of the ship with the air inside it is the same as the weight of the water it pushes aside.

When a submarine submerges, water rushes into tanks that were filled with air. The water pushes the air out of the tanks.

The water is like a ship's cargo. The submarine plus the water in its tanks is now heavier than the water the submarine pushes aside. The submarine submerges.

The submarine rises when air is pumped into the tanks. The air pushes the water out of the tanks.

Once again the submarine floats because it now weighs the same as the water it pushes aside.

Air helps things to float.

Here's a way to find out. Blow air into a balloon. Twist the neck of the balloon, and tie a string around it.

Push the balloon under water. Then let go. The balloon rises quickly. It may pop right out of the water. Then it floats on top.

Your own lungs are like tanks of air.
When you swim, the air in your lungs helps to keep you up.
When your lungs are full of air, you float. When your lungs do not have much air in them, you sink deeper in the water.

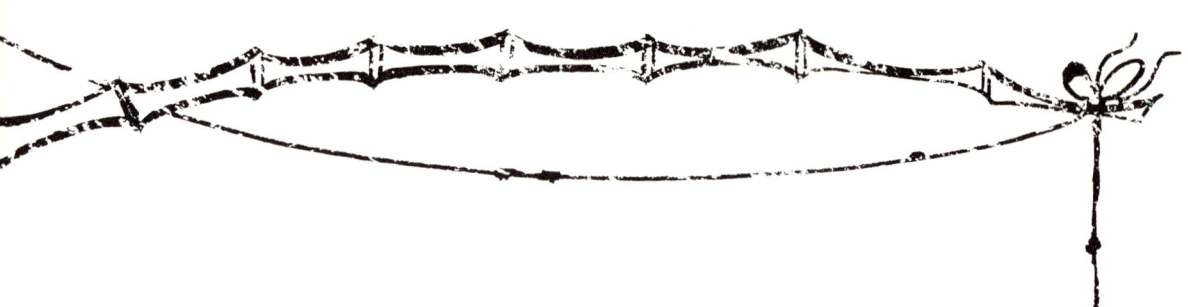

When you weigh more than the water you push aside, you sink. And other things do, also.
You know why things sink. And you know why things float.
You know that a bottle of air floats.
You know that a bottle of sand sinks.
Would a bottle of water float?
You can find out for yourself.

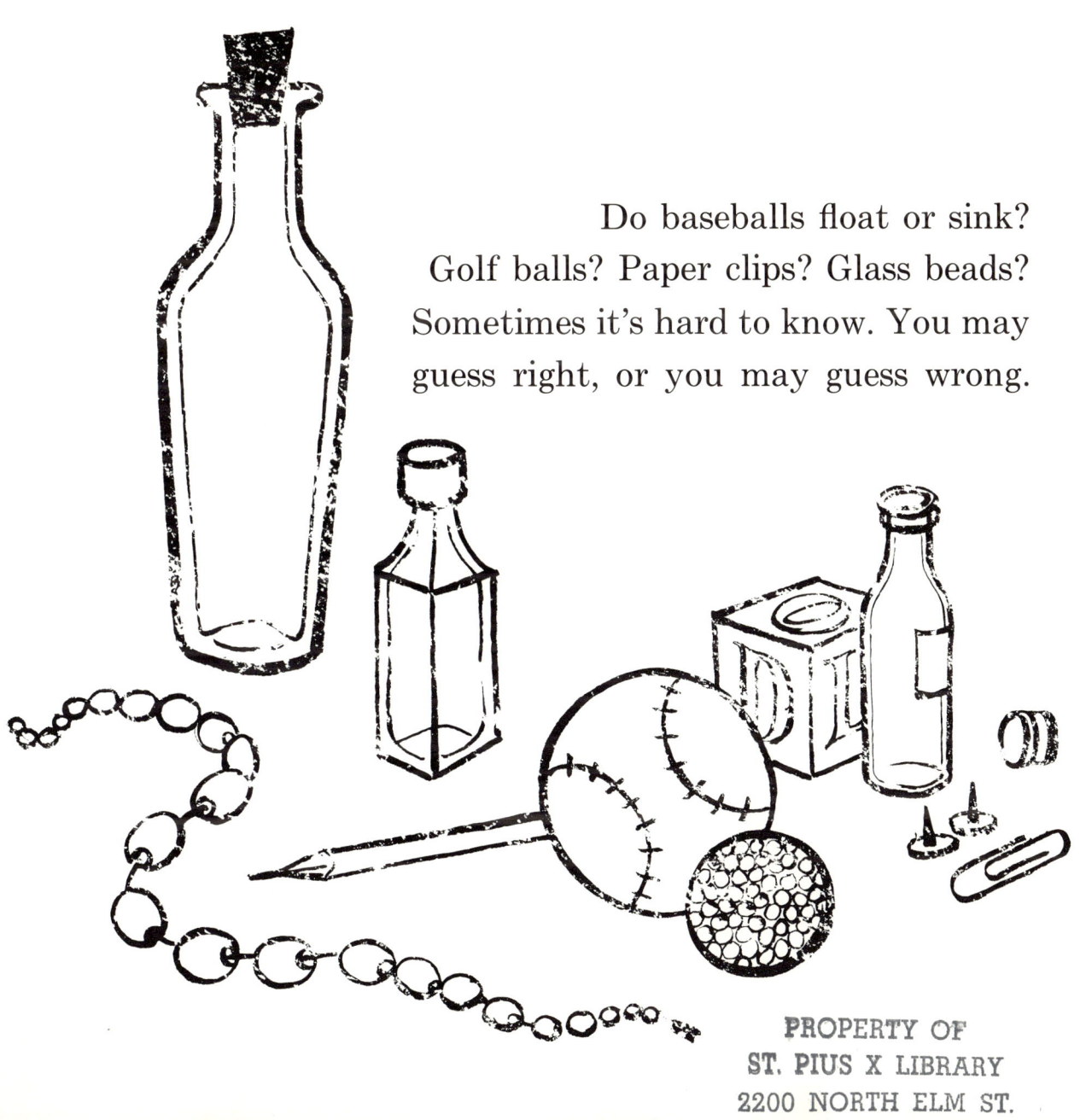

Do baseballs float or sink?
Golf balls? Paper clips? Glass beads?
Sometimes it's hard to know. You may
guess right, or you may guess wrong.

But you can find out.

ABOUT THE AUTHOR

Dr. Franklyn M. Branley is well known as the author of many excellent science books for young people of all ages. He is also co-editor of the Let's-Read-and-Find-Out Science Books.

Dr. Branley is Chairman and Astronomer of the American Museum-Hayden Planetarium in New York City.

He holds degrees from New York University, Columbia University, and from the New York State University College at New Paltz. He lives with his family in Woodcliff Lake, New Jersey.

ABOUT THE ILLUSTRATOR

Robert Galster has illustrated many books, designed book jackets and record-album covers, and painted murals. He is also well known for his Broadway-theater poster designs.

Mr. Galster was born in Illinois; he grew up in Ohio, and now lives in New York City.

DATE DUE

3197

E
532
Brd

Branley, Franklyn M.
  Floating and sinking
              B 22-495

**PROPERTY OF
ST. PIUS X LIBRARY
2200 NORTH ELM ST.**